The ESSENTIALS of
ALGEBRA &
TRIGONOMETRY II

P9-DBX-969

Staff of Research and Education Association,
Dr. M. Fogiel, Director

This book is a continuation of *"THE ESSENTIALS OF ALGEBRA AND TRIGONOMETRY I"* and begins with Chapter 12. It covers the usual course outline of Algebra and Trigonometry II. Earlier/basic topics are covered in *"THE ESSENTIALS OF ALGEBRA AND TRIGONOMETRY I"*.

Research and Education Association
61 Ethel Road West
Piscataway, New Jersey 08854

THE ESSENTIALS® OF ALGEBRA AND TRIGONOMETRY II

Printed in the United States of America

Library of Congress Catalog Card Number 87-61818

International Standard Book Number 0-87891-570-2

Revised Printing 1990

ESSENTIALS is a registered trademark of
Research and Education Association, Piscataway, New Jersey 08854

WHAT "THE ESSENTIALS" WILL DO FOR YOU

This book is a review and study guide. It is comprehensive and it is concise.

It helps in preparing for exams, in doing homework, and remains a handy reference source at all times.

It condenses the vast amount of detail characteristic of the subject matter and summarizes the **essentials** of the field.

It will thus save hours of study and preparation time.

The book provides quick access to the important facts, principles, theorems, concepts, and equations of the field.

Materials needed for exams, can be reviewed in summary form — eliminating the need to read and re-read many pages of textbook and class notes. The summaries will even tend to bring detail to mind that had been previously read or noted.

This "ESSENTIALS" book has been carefully prepared by educators and professionals and was subsequently reviewed by another group of editors to assure accuracy and maximum usefulness.

Dr. Max Fogiel
Program Director

CONTENTS

This book is a continuation of *"THE ESSENTIALS OF ALGEBRA AND TRIGONO-METRY I"* and begins with Chapter 12. It covers the usual course outline of Algebra and Trigonometry II. Earlier/basic topics are covered in *"THE ESSENTIALS OF ALGEBRA AND TRIGONOMETRY I"*.

v

CHAPTER 12

LOGARITHMS

12.1 LOGARITHMS

Logarithms were invented to make calculations with large numbers easier by expressing them in terms of their exponents. Given the exponential expression $x = b^y$, b ε N, b $\neq$ 0, we can rewrite it in its logarithmic form so that x is in terms of its exponent y, $y = \log_b x$. This expression is read as "y is equal to the logarithm to the base b of x."

If x, y and a are positive real numbers, a $\neq$ 1, and r is any real number, then

A) $\log_a(xy) = \log_a x + \log_a y$,

B) $\log_a\left(\dfrac{x}{y}\right) = \log_a x - \log_a y$,

C) $\log_a x^r = r \log_a x$.

Common logarithms are logarithms with a base of 10. We omit the base when working with base 10. That is

$$\log x = \log_{10} x.$$

The following formula will enable us to calculate non-common logarithms by using common logarithms.

$$\log_a b = \frac{\log_{10} b}{\log_{10} a}, \quad \text{where } a, b > 0.$$

For example, to find the value of $\log_7 3$, we use the above rule to obtain:

$$\log_7 3 = \frac{\log_{10} 3}{\log_{10} 7} \cong \frac{.4771}{.8451} \cong .5645$$

These values can be found by referring to log tables.

The antilogarithm is the number corresponding to a given logarithm. The cologarithm of a positive number is the logarithm of its reciprocal.

The common logarithm of any number is expressible as a combination of two parts:

A) the characteristic, which is the integral part;

B) the mantissa, which is the decimal part of the number.

To find the common logarithm of a positive number:

A) Express the number in scientific notation.

B) Determine the index of the number, which is the characteristic.

C) To find the mantissa, see a table of common logarithms of numbers.

E.g. Find the logarithm of 30,700.

Solution: First express 30,700 in scientific notation. 30,700 = 3.07×10^4. 4 is the characteristic. To find the mantissa, see a table of common logarithms of numbers. The mantissa is 4871. Thus log 30,700 = 4 + .4871 = 4.4871.

To find the antilogarithm:

A) Use the logarithm table to find the number that corresponds to that specific mantissa.

B) Rewrite that number in standard form.

C) Use the characteristic as the index for the number in standard form.

E.g. Find Antilog$_{10}$ 0.8762 - 2.

82

Solution: Let $N = $ Antilog$_{10}$ $0.8762 - 2$. The following relationship between log and antilog exists; $\log_{10}{}^x = a$ is the equivalent of $x = $ antilog$_{10}$ a. Therefore,

$$\log_{10} N = 0.8762 - 2.$$

The characteristic is -2. The mantissa is 0.8762. The number that corresponds to this mantissa is 7.52. This number is found from a table of common logarithms, base 10. Therefore,

$$N = 7.52 \times 10^{-2}$$

$$= 7.52 \times \left(\frac{1}{10^2} \right)$$

$$= 7.52 \times \left(\frac{1}{100} \right)$$

$$= 7.52(.01)$$

$$N = 0.0752.$$

Therefore, $N = $ Antilog$_{10}$ $0.8762 - 2 = 0.0752$.

12.2 LOGARITHMIC, EXPONENTIAL AND POWER FUNCTIONS

The function $f(x) = x^n$ is called a power function in x. An exponential function in x is of the form $f(x) = a^x$. A logarithmic function in x is of the form $f(x) = \log_a x$.

An equation involving one or more unknowns in an exponent is called an exponential equation. $2^x + 6x + 8^{y+7} = y$ is an exponential equation in two unknowns.

The following are some examples of the graphs of power, exponential and logarithmic functions.

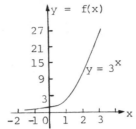

Fig. 12.1 Exponential function

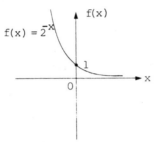

Fig. 12.2 Exponential function

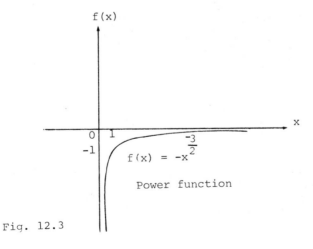

Power function

Fig. 12.3

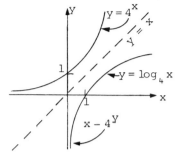

Fig. 12.4 Exponential, Power, and
Logarithmic Functions

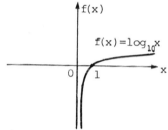

Fig. 12.5 Logarithmic function

CHAPTER 13

SEQUENCES AND SERIES

13.1 SEQUENCES

A set of numbers $u_1, u_2, u_3, \ldots$ in a definite order of arrangement and formed according to a definite rule is called a sequence. Each number in the sequence is called a term of the sequence. If the number of terms is finite, it is called a finite sequence, otherwise it is called an infinite sequence.

A general term n can be obtained by applying a general law of formation, by which any term in the sequence can be obtained.

An arithmetic progression (A.P.) is a sequence of numbers where each term excluding the first is obtained from the preceding one by adding a fixed quantity to it. This constant amount is called the common difference.

Let   a = first term of progression

l = last term

d = common difference

k = number of terms

S_n = sum of first n terms, $n \leq k$

then

$$l = a + (k - 1)d$$

$$S_n = \frac{k}{2}(a + l) = \frac{k}{2}[2a + (k - 1)d]$$

In general, to find the common difference in a given arithmetic progression, simply subtract any term from its successor.

A single term between two given terms a and b is called the arithmetic mean, M, between these two terms and is given by:

$$M = \frac{a + b}{2}$$

E.g. Insert 4 arithmetic means between 1 and 36.

Solution: The terms between any two given terms of a progression are called the means between these two terms. Inserting 4 arithmetic means between 1 and 36 requires an arithmetic progression (A.P.) of the form 1, ___, ___, ___, ___, 36. Using the formula, $l = a + (k - 1)d$, for the nth term, we can determine d. Knowing the common difference, d, we can obtain the means by adding d to each preceding number after the first.

a = 1, l = 36, and k = 6 since there will be six terms

$l = a + (k - 1)d$

$36 = 1 + 5d$

$5d = 35$

$d = 7$

The arithmetic means are: 1+7, (1+7)+7, (1+7+7)+7, (1+7+7+7)+7; that is, 8, 15, 22, and 29. The arithmetic progression is 1, 8, 15, 22, 29, 36.

A geometric progression (G.P.) is a sequence of numbers each of which, after the first, is obtained by multiplying the preceding number by a constant number called the common ratio, r.

Let a = first term

r = common ratio

k = number of terms

l = last term

S_n = sum of first n terms

then

$$l = ar^{k-1}$$

$$S_n = \frac{a(r^k - 1)}{r - 1} = \frac{rl - a}{r - 1}$$

The sum to infinity (S_∞) of any geometric progression is

$$S_\infty = \frac{a}{1 - r} \quad \text{if} \quad |r| < 1$$

To find the common ratio in a given geometric progression, divide any term by its predecessor.

A single term between two given terms of a geometric progression is called the geometric mean between the two terms. The geometric mean is denoted by G.

If G is the geometric mean between the terms a and b, then $\frac{a}{G} = \frac{G}{b}$ which implies:

$$G^2 = ab \quad \text{or} \quad G = \pm\sqrt{ab}$$

A harmonic progression (H.P.) is a sequence of numbers whose reciprocals are an arithmetic progression. The terms between any two given terms of a harmonic progression are called the harmonic means between these two terms.

Let   a = first term

       l = last term

       k = number of terms

       d = common difference

then

$$l = \frac{1}{a + (n - 1)d}$$

E.g.   If a^2, b^2, c^2 are in arithmetic progression, show that $b + c$, $c + a$, $a + b$ are in harmonic progression.

Solution: We are given that a^2, b^2, c^2 are in arithmetic progression. By this we mean that each new term is obtained by adding a constant to the preceding term.

By adding (ab + ac + bc) to each term, we see that

$a^2 + (ab + ac + bc)$, $b^2 + (ab + ac + bc)$,

$c^2 + (ab + ac + bc)$

are also in arithmetic progression. These three terms can be rewritten as

$a^2 + ab + ac + bc$, $b^2 + bc + ab + ac$,
$c^2 + ac + bc + ab$

Notice that:

$a^2 + ab + ac + bc = (a + b)(a + c)$

$b^2 + bc + ab + ac = (b + c)(b + a)$

$c^2 + ac + bc + ab = (c + a)(c + b)$

Therefore, the three terms can be rewritten as:

$(a + b)(a + c)$, $(b + c)(b + a)$, $(c + a)(c + b)$,

which are also in arithmetic progression.

Now, dividing each term by $(a + b)(b + c)(c + a)$, we obtain:

$$\frac{1}{b + c}, \quad \frac{1}{c + a}, \quad \frac{1}{a + b},$$

which are in arithmetic progression.

Recall that a sequence of numbers whose reciprocals form an arithmetic progression, is called a harmonic progression. Thus, since $\frac{1}{b + c}$, $\frac{1}{c + a}$, $\frac{1}{a + b}$ is an arithmetic progression, $b + c$, $c + a$, $a + b$ are in harmonic progression.

13.2 SERIES

A series is defined as the sum of the terms of a

sequence $u_1 + u_2 + u_3 + \ldots + u_n$.

The terms $u_1, u_2, u_3, \ldots, u_n$ are called the first, second, third and nth terms of the series. If the series has a finite number of terms it is called a finite series, otherwise it is called an infinite series.

Finite series

$$\sum_{i=1}^{n} u_i = u_1 + u_2 + u_3 + \ldots + u_n$$

Infinite series

$$\sum_{i=1}^{\infty} u_i = u_1 + u_2 + u_3 + \ldots$$

The general or nth term of a series is an expression which indicates the law of formation of the terms.

E.g. Determine the general term of the series:

$$\frac{1}{5^3} + \frac{3}{5^5} + \frac{5}{5^7} + \frac{7}{5^9} + \frac{9}{5^{11}} + \ldots$$

<u>Solution</u>: The numerators of the terms in the series are consecutive odd numbers beginning with 1. An odd number can be represented by $2n - 1$.

In the denominators, the base is always 5, and the power is a consecutive odd integer beginning with 3.

The general term can therefore be expressed by

$$\frac{2n - 1}{5^{2n+1}}, \text{ the series by } \sum_{i=1}^{\infty} \frac{2i - 1}{5^{2i+1}}$$

and the series is generated by replacing i with i = 1, 2, 3, 4, ...

Let $s_n = u_1 + u_2 + \ldots + u_n$ be the sum of the first n terms of the infinite series $u_1 + u_2 + u_3 + \ldots$. The terms of the sequence $s_1, s_2, s_3 \ldots$ are called the partial sums of the series.

If the values of $s_1, s_2, \ldots, s_n$ never become greater or equal than a certain value s, no matter how big n is, and approach s as n increases, the sums are said to have a limit, and this is represented by:

$$\lim_{n \to \infty} s_n = s.$$

If $\lim_{n \to \infty} s_n = s$ is a finite number the series $u_1 + u_2 + u_3 + \ldots$ is convergent and s is called the sum of the infinite series.

A series which is not convergent is said to be divergent.

An alternating series is one whose terms are alternately positive or negative. An alternating series converges if:

A) After a certain number of terms the absolute value of a certain term is less than that of the preceding term.

B) The nth term has a limit of zero as $n \to \infty$.

A series is said to be absolutely convergent if the series formed by taking absolute values of the terms converges. A convergent series which is not absolutely convergent is conditionally convergent.

The terms of an absolutely convergent series may be arranged in any order and not affect the convergence.

A series of the form $c_0 + c_1 x + c_2 x^2 + \ldots$ where the coefficients $c_0, c_1, c_2, \ldots$ are called constants is called a power series in x. It is denoted by

$$\sum_{n=0}^{\infty} c_n x^n$$

The set of values of x for which a power series converges is called its interval of convergence.

A series of the type $\dfrac{1}{1^p} + \dfrac{1}{2^p} + \dfrac{1}{3^p}$ where p is a constant is known as a p series and is denoted by

$$\sum_{n=1}^{\infty} \frac{1}{n^p}.$$

The p series converges if p ≤ 1.

The following methods are used to test convergence of series:

A) Comparison test for convergence of series of positive terms.

If each term of a given series of positive terms is less than or equal to the corresponding term of a known convergent series, then the given series converges.

If each term of a given series of positive terms is greater than or equal to the corresponding term of a known divergent series, then the given series diverges.

E.g.   Establish the convergence or divergence of the series:

$$\frac{1}{1 + \sqrt{1}} + \frac{1}{1 + \sqrt{2}} + \frac{1}{1 + \sqrt{3}} + \frac{1}{1 + \sqrt{4}} + \ldots$$

Solution: To establish the convergence or divergence of the given series we first determine the nth term of the series. By studying the law of formation of the terms of the series we find the nth term to be $\frac{1}{1 + \sqrt{n}}$.

To determine whether this series is convergent or divergent we use the comparison test. We choose $\frac{1}{n}$, which is a known divergent series since it is a p-series, $\frac{1}{n^p}$, with p = 1. If we can show $\frac{1}{1 + \sqrt{n}} > \frac{1}{n}$, then $\frac{1}{1 + \sqrt{n}}$ is divergent. But we can see this is true, since $1 + \sqrt{n} < n$ for n > 2. Therefore the given series is divergent.

B) Ratio Test

For a given series $s_1 + s_2 + s_3 + \ldots$ it is possible to conclude it is:

$$\text{Convergent if:} \quad \lim_{n \to \infty} \left| \frac{s_{n+1}}{s_n} \right| = L < 1$$

$$\text{Divergent if:} \quad \lim_{n \to \infty} \left| \frac{s_{n+1}}{s_n} \right| = L > 1$$

If $\lim_{n \to \infty} \left| \dfrac{s_{n+1}}{s_n} \right| = L = 1$ the ratio test is not decisive, it fails to establish convergence or divergence.

CHAPTER 14

PERMUTATIONS, COMBINATIONS AND PROBABILITY

14.1 PERMUTATION

A permutation is an arrangement of all or part of a number of objects in any order.

The symbol $_bP_a$ represents the number of permutations of "b" objects taken "a" at a time.

$$_bP_a = \frac{b!}{(b - a)!}$$

E.g. $_9P_4 = \frac{9!}{(9 - 4)!} = \frac{9!}{5!} = \frac{9 \times 8 \times 7 \times 6 \times 5 \times 4 \times 3 \times 2 \times 1}{5 \times 4 \times 3 \times 2 \times 1}$

$= 3,024$

14.2 COMBINATION

A combination is a grouping of all or part of a number of objects without regard to the order of the arrangement of the selected objects.

The symbol $_bC_a$, represents the combination of "b" different objects taken "a" at a time without regard to the order of the arrangement of the selected objects.

$$_bC_a = \frac{b!}{a!(b-a)!} = \frac{_bP_a}{a!}$$

Note that $_bC_a$ can also be written as $\begin{pmatrix} b \\ a \end{pmatrix}$, read "b choose a."

E.g. $\quad _9C_4 = \frac{_9P_4}{4!} = \frac{3024}{24} = 126$

Also note that $_bC_a = {_bC_{b-a}}$.

Given n objects, if we take them 1 at a time, 2 at a time, 3 at a time,...,n at a time and add all these combinations we will obtain C which is

$$C = 2^n - 1.$$

14.3 PROBABILITY

If an event can occur in k ways and fail to occur in m ways and all of these (k + m) ways are assumed equally likely, then the probability of the event to occur (success) is $P(s) = \frac{k}{k+m}$ and the probability of the event not to occur (failure) is $P(f) = \frac{m}{k+m}$.

95

Law of Total Probability

$$P(s) + P(f) = 1$$

This states an event will either occur or it won't.

Two or more events are said to be independent if the occurrence of any of them does not affect the probability of occurrence of any of the others.

The probability that two or more independent events will occur one after the other equals the product of their separate probabilities.

Two or more events are said to be dependent if the occurrence of one of the events affects the probability of occurrence of the others.

Two or more events are said to be mutually exclusive if the occurrence of any of them excludes the occurrence of the others.

The probability of occurrence of one, two or more mutually exclusive events is the sum of the probabilities of the individual events.

Let p be the probability that a given event will occur in any single trial and q be the probability that it will fail to occur in any single trial. Then the probability that the event will occur exactly k times in n independent trials is $_nC_k p^k q^{n-k}$.

The probability that a given event will occur <u>at least</u> m times in n independent trials is given below:

$$P(s \geq m) = {}_nC_m p^m q^{n-m} + {}_nC_{m+1} p^{m+1} q^{n-(m+1)}$$

$$+ \ _{n}C^{m+2}p^{m+2}q^{n-(m+2)} + \ldots + \ _{n}C_{n}p^{n}q^{0}.$$

where q = 1 - p.

Note that the probability that an event will occur at least m times in n trials, corresponds to the sum of the probabilities of the event to occur m, m+1, m+2, ..., n times.

The probability of any event to occur is between zero and one inclusive.

$$0 \leq P(s) \leq 1$$

P(s) = 0 indicates the event will not occur.

P(s) = 1 indicates the event will definitely occur.

E.g. A die is tossed five times. What is the probability that a one will appear at least twice?

To find the probability that a one will occur at least twice, find the probability that it will occur twice, three times, four times, and five times. The sum of these probabilities will be that a one will happen at least twice. P(s) = probability that a one will occur in a given trial.

$$P(s) = \frac{\text{number of ways to obtain a one}}{\text{number of ways to obtain any face of a die}}$$

An experiment can only succeed or fail, hence the probability of success, p(s), plus the probability of failure, p(f), is one; p(s) + p(f) = 1. Then p(f) = 1 - p(s) = 1 - $\frac{1}{6}$ = $\frac{5}{6}$.

$$P(s \geq 2) = \ _{5}C_{2}\left(\frac{1}{6}\right)^{2}\left(\frac{5}{6}\right)^{3} + \ _{5}C_{3}\left(\frac{1}{6}\right)^{3}\left(\frac{5}{6}\right)^{2} + \ _{5}C_{4}\left(\frac{1}{6}\right)^{4}\left(\frac{5}{6}\right)^{1}$$

$$+ \ _{5}C_{5}\left(\frac{1}{6}\right)^{5}\left(\frac{5}{6}\right)^{0}$$

Then,

$$_{5}C_{2}\left(\frac{1}{6}\right)^{2}\left(\frac{5}{6}\right)^{3} + \ _{5}C_{3}\left(\frac{1}{6}\right)^{3}\left(\frac{5}{6}\right)^{2} + \ _{5}C_{4}\left(\frac{1}{6}\right)^{4}\left(\frac{5}{6}\right)^{1}$$

$$+ \; _5C_5 \left(\frac{1}{6}\right)^5 \left(\frac{5}{6}\right)^0$$

$$= \frac{5!}{2!3!} \left(\frac{125}{6^5}\right) + \frac{5!}{2!3!} \left(\frac{25}{6^5}\right) + \frac{5!}{4!1!} \left(\frac{5}{6^5}\right) + \frac{5!}{5!0!} \left(\frac{1}{6^5}\right)$$

$$= \frac{5 \times 4 \times 3!}{2 \times 3!} \left(\frac{125}{6^5}\right) + \frac{5 \times 4 \times 3!}{2 \times 3!} \left(\frac{25}{6^5}\right) + \frac{5 \times 4!}{4!1!} \left(\frac{5}{6^5}\right) + \frac{1}{6^5}$$

$$= 10 \left(\frac{125}{6^5}\right) + 10 \left(\frac{25}{6^5}\right) + 5 \left(\frac{5}{6^5}\right) + \frac{1}{6^5}$$

$$= \frac{1250 + 250 + 25 + 1}{6^5} = \frac{1526}{7776} = \frac{763}{3888}$$

Therefore, the probability that a one will appear at least twice is $\frac{763}{3888}$.

CHAPTER 15

VECTORS, MATRICES, DETERMINANTS AND SYSTEMS OF EQUATIONS

15.1 VECTORS

A vector is a quantity having both magnitude and direction, such as displacement, velocity, force, or acceleration.

A scalar is a quantity having magnitude but no direction, e.g. mass, length, time, temperature, or any real number.

A vector in one-, two- or three-dimensional space can be represented by an arrow; the length of the arrow indicates the magnitude of the vector while its direction indicates the direction of the vector.

Let a vector extend from a point P to a point Q. Then the vector is denoted by $\vec{PQ}$, and the magnitude is denoted by $|\vec{PQ}|$.

Two vectors are said to be equal if they have the same magnitude and direction.

If $V(X_0, Y_0, Z_0)$ represents a point in three-dimensional space, and O represents the origin $(0,0,0)$, then $\vec{OV}$ is called the position vector of the point $V(X_0, Y_0, Z_0)$. This vector can be denoted by $\vec{P} = <X_0, Y_0, Z_0>$ where X_0, Y_0 and Z_0 are called the components of $\vec{P}$. The magnitude of $\vec{P}$ is given by

$$|\vec{P}| = \sqrt{X_0^2 + Y_0^2 + Z_0^2}.$$

If $A(X_1, Y_1, Z_1)$ and $B(X_2, Y_2, Z_2)$ are two points in three-dimensional space (3-space), then the vector $\vec{AB}$ is given by

$$\vec{AB} = <X_2 - X_1,\quad Y_2 - Y_1,\quad Z_2 - Z_1>$$

If $\vec{p}$, $\vec{q}$ and $\vec{r}$ are any vectors in three-space, and a, b and z are scalars, $z = 0$; let $\vec{n}$ be the null vector, i.e. $|\vec{n}| = 0$ then:

A) $\vec{p} + \vec{q} = \vec{q} + \vec{p}$

B) $\vec{p} - \vec{q} = \vec{p} + (-\vec{q}) = (-\vec{q}) + \vec{p}$

C) $\vec{p} + (\vec{q} + \vec{r}) = (\vec{p} + \vec{q}) + \vec{r}$

D) $\vec{p} + (-\vec{p}) = \vec{n}$

E) $\vec{p} + \vec{n} = \vec{p}$

F) $a(\vec{p} + \vec{q}) = a\vec{p} + a\vec{q} = \vec{p}a + \vec{q}a = (\vec{p} + \vec{q})a$

G) $(a + b)\vec{p} = a\vec{p} + b\vec{p} = \vec{p}a + \vec{p}b = \vec{p}(a + b)$

H) $a \cdot b\vec{p} = a(b\vec{p}) = b(a\vec{p}) = ba \cdot \vec{p} = \vec{p}(ab)$

I) $z \cdot \vec{p} = \vec{n}$

$-\vec{p}$ has the same magnitude as $\vec{p}$ but is opposite in direction.

Algebraically, the sum of two vectors is found by adding the corresponding coordinates of the vectors.

Similarly, the difference of two vectors is found by subtracting corresponding coordinates.

e.g. $\vec{a} = <2,1>$, $\vec{b} = <-3,5>$

so $\vec{a} + \vec{b} = <2 + (-3), 1 + 5> = <-1,6>$

    The sum and difference of two vectors can also be found graphically, by the triangle and parallelogram laws of vector addition. The figures shown below are self-explanatory.

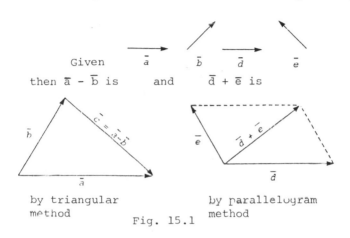

Fig. 15.1

    Finally, in a rectangular coordinate system in three dimensions a vector $\vec{P} = <x_1, y_1, z_1>$ may also be represented as

$$\vec{p} = x_1\vec{i} + y_1\vec{j} + z_1\vec{k}$$

where $\vec{i} = <1,0,0>$, $\vec{j} = <0,1,0>$ and $\vec{k} = <0,0,1>$. They are called unit vectors.

15.2 MATRICES

    A matrix is a rectangular array of numbers, or variables.

Examples:

$$\begin{pmatrix} 2 & 2 & 1 \\ 1 & 0 & 4 \end{pmatrix}, \quad \begin{pmatrix} a_{11} & a_{12} & a_{13} \\ a_{21} & a_{22} & a_{23} \\ a_{31} & a_{32} & a_{33} \\ a_{41} & a_{42} & a_{43} \end{pmatrix}, \quad \begin{pmatrix} 8 & 4 & \frac{5}{3} & 7 \\ 9 & 0 & 1 & -1 \\ 2 & 3 & 4 & 5 \end{pmatrix}$$

Each number in a given matrix is called an element of the matrix. An element may be denoted by a_{ij} where i indicates the row containing the element and j indicates the column containing the element.

A matrix of m rows and n columns is said to be a matrix of order m by n (m × n). An n × n matrix is called a square matrix. The principal diagonal of a square matrix is the diagonal containing the elements from the upper left-hand corner to the lower right-hand corner.

The identity matrix I of order n × n is a matrix whose elements are all zeros except elements on the principal diagonal which are ones.

$$I_{3 \times 3} = \begin{array}{l} \\ \text{R} \\ \text{o} \\ \text{w} \\ \text{s} \end{array} \begin{pmatrix} 1 & 0 & 0 \\ 0 & 1 & 0 \\ 0 & 0 & 1 \end{pmatrix} \xrightarrow{\text{Columns}}$$

A square matrix is said to be diagonal if entries off its principal diagonal are zero and entries on the principal diagonal are any real numbers.

A matrix is in triangular form if all the elements below the principal diagonal are zeros.

Given a system of equations of the form,

$$a_{11}x_1 + a_{12}x_2 + a_{13}x_3 + \ldots + a_{1n}x_n = c_1$$
$$a_{21}x_1 + a_{22}x_2 + a_{23}x_3 + \ldots + a_{2n}x_n = c_2$$
$$\vdots \qquad \vdots \qquad \vdots \qquad \qquad \vdots \qquad \vdots$$
$$a_{m1}x_1 + a_{m2}x_2 + a_{m3}x_3 + \ldots + a_{mn}x_n = c_m$$

the augmented matrix of this system of equations is

$$\left(\begin{array}{ccccc|c} a_{11} & a_{12} & a_{13} & \cdots & a_{1n} & c_1 \\ a_{21} & a_{22} & a_{23} & \cdots & a_{2n} & c_2 \\ \vdots & & & & \vdots & \vdots \\ a_{m1} & a_{m2} & a_{m3} & \cdots & a_{mn} & c_m \end{array}\right)$$

and the coefficient matrix of this system is

$$\left(\begin{array}{ccccc} a_{11} & a_{12} & a_{13} & \cdots & a_{1n} \\ a_{21} & a_{22} & a_{23} & \cdots & a_{2n} \\ \vdots & \vdots & \vdots & & \vdots \\ a_{m1} & a_{m2} & a_{m3} & \cdots & a_{mn} \end{array}\right)$$

Any of the following operations on a matrix is said to be an elementary row operation.

A) Interchange of two rows of a matrix.

B) Multiplication of each element of a row by the same non-zero constant.

C) Addition of the elements of a row multiplied by a non-zero constant to the corresponding elements of another row.

Two matrices are equal if and only if they are of the same order and have the same entries in each position.

To add or subtract matrices is to add or subtract the corresponding entries of the matrices.

Example:

$$\begin{pmatrix} 9 & 1 & 2 \\ 0 & 1 & 0 \end{pmatrix} + \begin{pmatrix} 0 & 0 & 0 \\ 0 & 1 & 1 \end{pmatrix}$$

$$= \begin{pmatrix} 9+0 & 1+0 & 2+0 \\ 0+0 & 1+1 & 0+1 \end{pmatrix} = \begin{pmatrix} 9 & 1 & 2 \\ 0 & 2 & 1 \end{pmatrix}$$

Product of two matrices. The product C of two matrices, $A = [a_{ij}]_{m \times n}$ and $B = [b_{ij}]_{n \times q}$, which are conformable for multiplication in the order AB is defined by

$$A \times B = C = [c_{ij}]_{m \times q}$$

where

$$c_{ij} = a_{i1}b_{1j} + a_{i2}b_{2j} + \ldots + a_{in}b_{nj} = \sum_{k=1}^{n} a_{ik}b_{kj}$$

In other words, we say that the (i,j) element c_{ij} of the product matrix $C = A \times B$ is the sum of the products of the elements in the ith row of A and the corresponding elements in the jth column of B.

Example:

$$A = \begin{bmatrix} 3 & -5 \\ 7 & 0 \end{bmatrix} \qquad B = \begin{bmatrix} 2 & 4 \\ -8 & 9 \end{bmatrix}$$

Then

$$AB = \begin{bmatrix} 3 & -5 \\ 7 & 0 \end{bmatrix}\begin{bmatrix} 2 & 4 \\ -8 & 9 \end{bmatrix} = \begin{bmatrix} 3(2)+(-5)(-8) & 3(4)+(-5)9 \\ 7(2)+(0)(-8) & 7(4)+(0)9 \end{bmatrix}$$

$$= \begin{bmatrix} 46 & -33 \\ 14 & 28 \end{bmatrix}$$

$$BA = \begin{bmatrix} 2 & 4 \\ -8 & 9 \end{bmatrix}\begin{bmatrix} 3 & -5 \\ 7 & 0 \end{bmatrix} = \begin{bmatrix} 2(3)+4(7) & 2(-5)+4(0) \\ -8(3)+9(7) & (-8)(-5)+9(0) \end{bmatrix}$$

$$= \begin{bmatrix} 34 & -10 \\ 39 & 40 \end{bmatrix}$$

In this case $AB \neq BA$. Hence, matrix multiplication is not commutative.

Transpose of a matrix. The matrix A^T of order $n \times m$, obtained by interchanging the rows and columns of an $m \times n$ matrix A, is defined as the transpose of A. In symbolic terms, if $A = [a_{ij}]_{m \times n}$, then $A^T = [a_{ij}]_{n \times m}$.

Example:

If
$$A = \begin{bmatrix} a & b \\ c & d \end{bmatrix}$$

then

$$A^T = \begin{bmatrix} a & c \\ b & d \end{bmatrix}$$

If A, B and C are any three $m \times n$ matrices, and a, b, c are real constants, let Z denote the zero matrix then

A) $(A \pm B) \pm C = A \pm (B \pm C)$

B) $A + (-A) = Z$, where Z is the zero matrix with all entries equal to zero

C) $A + Z = A$

D) $(a + b)A = aA + bA$

E) $a(A + B) = aA + aB$

F) $(ab)A = b(aA) = a(bA)$

G) $I_{m \times m}A = A$, where $I_{m \times m}$ is the identity matrix.

15.3 DETERMINANTS

For a 2×2 matrix, the determinant is given by:

$$\begin{vmatrix} a_1 & b_1 \\ a_2 & b_2 \end{vmatrix} = a_1 b_2 - b_1 a_2$$

For a 3×3 matrix, the determinant is given by:

$$\begin{vmatrix} a_1 & b_1 & c_1 \\ a_2 & b_2 & c_2 \\ a_3 & b_3 & c_3 \end{vmatrix} = a_1 b_2 c_3 + b_1 c_2 a_3 + c_1 a_2 b_3 - c_1 b_2 a_3 - a_1 c_2 b_3 - b_1 a_2 c_3$$

Another way to find the determinant of a 3×3 matrix is as follows:

A) Rewrite the first two columns on the right of the determinant.

B) Compute the products of the three diagonals running from left to right; prefix each term by a positive sign.

C) Compute the products of the three diagonals running from right to left; prefix each term by a negative sign.

D) The sum of the six products is the value of the determinant.

The following figure may be helpful for remembering this method.

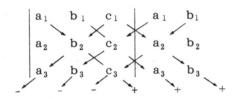

A third method for finding the determinant of a 3×3 matrix is the following:

$$\begin{pmatrix} a_1 & a_2 & a_3 \\ b_1 & b_2 & b_3 \\ c_1 & c_2 & c_3 \end{pmatrix}$$

$$\det \begin{pmatrix} a_1 & a_2 & a_3 \\ b_1 & b_2 & b_3 \\ c_1 & c_2 & c_3 \end{pmatrix} = a_1 \begin{vmatrix} b_2 & b_3 \\ c_2 & c_3 \end{vmatrix} - a_2 \begin{vmatrix} b_1 & b_3 \\ c_1 & c_3 \end{vmatrix} + a_3 \begin{vmatrix} b_1 & b_2 \\ c_1 & c_2 \end{vmatrix}$$

Note that we start with a positive sign, and alternate signs after.

Properties of Determinants

A) If matrix A is obtained from matrix B by interchanging any two rows or columns of B, then $\det A = -\det B$.

B) A determinant does not change its value if a linear

combination of other rows (columns) is added to any given row (column).

C) If A is obtained from B by multiplying one row of B by a non-zero constant a, then

$$\det A = a \det B.$$

D) If two rows or columns of a matrix A are identical, then

$$\det A = 0.$$

E) If I is the identity matrix, then

$$\det I = 1.$$

F) If an entire row (or a column) of a matrix consists of zero elements only, the determinant of this matrix is zero.

G) If a square matrix is in triangular form, then the determinant of this matrix is the product of all the numbers on the principal diagonal.

15.4 SYSTEM OF EQUATIONS

Consider a system of n linear equations in n unknowns:

$$a_{11}x_1 + a_{12}x_2 + \dots + a_{1n}x_n = b_1$$
$$a_{21}x_1 + a_{22}x_2 + \dots + a_{2n}x_n = b_2$$
$$\cdot \quad \cdot \quad \cdot \quad \cdot \quad \cdot \quad \cdot \quad \cdot \quad \cdot \quad \cdot \quad \cdot \quad \cdot \quad \cdot$$
$$a_{n1}x_1 + a_{n2}x_2 + \dots + a_{nn}x_n = b_n.$$

Transforming the above equations into matrix notation, we get

$$\begin{bmatrix} a_{11} & a_{12} & \cdots & a_{1}n \\ a_{21} & a_{22} & \cdots & a_{2}n \\ \vdots & & \vdots & \\ a_{n1} & a_{n2} & \cdots & a_{nn} \end{bmatrix} \begin{bmatrix} x_1 \\ x_2 \\ \vdots \\ x_n \end{bmatrix} = \begin{bmatrix} b_1 \\ b_2 \\ \vdots \\ b_n \end{bmatrix}$$

or, $AX = B$.

Let A be an $n \times n$ matrix over the field F such that det $A \neq 0$. If $b_1, b_2, \ldots, b_n$ are any scalars in F, the unique solution of the system of equations $AX = B$ is given by:

$$x_i = \frac{\det A_i}{\det A} \qquad i = 1, 2, \ldots, n,$$

where A_i is the $n \times n$ matrix obtained from A by replacing the ith column of A by the column vector.

$$\begin{bmatrix} b_1 \\ b_2 \\ \vdots \\ b_n \end{bmatrix}$$

The above theorem is known as "Cramer's Rule" for solving systems of linear equations. Cramer's Rule applies only to systems of n linear equations in n unknowns with non-zero determinants.

Ex: Solve the given system of equations, by method of determinants

$$3x - 5y = 4$$

$$7x + 4y = 25.$$

Solution: The equations, as given, are in standard form for applying Cramer's rule.   Therefore,

$$x = \frac{\begin{vmatrix} 4 & -5 \\ 25 & 4 \end{vmatrix}}{\begin{vmatrix} 3 & -5 \\ 7 & 4 \end{vmatrix}} = \frac{4 \cdot 4 - 25(-5)}{3 \cdot 4 - 7(-5)} = \frac{16 + 125}{12 + 35} = \frac{141}{47} = 3.$$

$$y = \frac{\begin{vmatrix} 3 & 4 \\ 7 & 25 \end{vmatrix}}{\begin{vmatrix} 3 & -5 \\ 7 & 4 \end{vmatrix}} = \frac{3 \cdot 25 - 7 \cdot 4}{47} = \frac{75 - 28}{47} = \frac{47}{47} = 1$$

This process always yields a unique solution unless the denominator determinant is equal to zero.

A system of linear equations is said to be in echelon (or triangular) form if the coefficient matrix of the system is in triangular form.

Example:
$$2x_1 + 4x_2 + x_3 = 0$$
$$x_2 + x_3 = 5$$
$$x_3 = 4$$

A system of k linear equations in k unknowns is determinative if and only if the determinant of the coefficient matrix of this system is not zero.

If the system is determinative then it is consistent, i.e. contains no contradictions like:

$$x + y = 0$$
$$x + y = 1.$$

CHAPTER 16

MATHEMATICAL INDUCTION AND THE BINOMIAL THEOREM

16.1 MATHEMATICAL INDUCTION

Mathematical induction is a method of proof. The steps are:

A) Verification of the proposed formula or theorem for the smallest value of n.

B) Assume that the theorem is true for n = k then prove that it is true for n = k + 1.

C) Conclude that the proposed theorem holds true for all values of n.

Example: Prove by mathematical induction that

$$1 + 7 + 13 + \ldots + (6n - 5) = n(3n - 2).$$

Solution: (A) The proposed formula is true for n = 1, since $1 = 1(3 - 2)$.

(B) Assume the formula to be true for n = k, a positive integer; that is, assume

a) $1 + 7 + 13 + \ldots + (6k - 5) = k(3k - 2)$.

Under this assumption we wish to show that

b) $1 + 7 + 13 + \ldots + (6k - 5) + [6(k + 1) - 5] = (k + 1)[3(k + 1) - 2]$, which is equivalent to when $(6k + 1)$ is added to both sides of a), we have on the right $k(3k - 2) + (6k + 1) = 3k^2 + 4k + 1 = (k + 1)(3k + 1)$; hence, if the formula is true for $n = k$ it is true for $n = k + 1$.

(C) Since the formula is true for $n = k = 1$ (Step A), it is true for $n = k + 1 = 2$; being true for $n = k = 2$ it is true for $n = k + 1 = 3$; and so on, for every positive integral value of n.

Definition: The expression n! is read "n factorial" and is the product of all consecutive integers from n down to 1.

Examples: $3! = 3 \cdot 2 \cdot 1$

$7! = 7 \cdot 6 \cdot 5 \cdot 4 \cdot 3 \cdot 2 \cdot 1$

Note: $0! = 1$ by definition.

16.2 BINOMIAL THEOREM

The following equation is the binomial theorem or binomial expansion:

$$(x + y)^n = \sum_{k=0}^{n} \binom{n}{k} x^{n-k} y^k$$

$$= \binom{n}{0} x^n + \binom{n}{1} x^{n-1} y + \binom{n}{2} x^{n-2} y^2 +$$

$$\ldots + \binom{n}{k} x^{n-k} y^k + \binom{n}{n} y^n$$

Ex.   Find the expansion of $(a - 2x)^7$.

Solution: Use the binomial formula:

$$(u + v)^n = u^n + nu^{n-1}v + \frac{n(n - 1)}{2} u^{n-2}v^2$$

$$+ \frac{n(n - 1)(n - 2)}{2 \cdot 3} u^{n-3}v^3 + \ldots + v^n$$

and substitute a for u and (-2x) for v and 7 for n to obtain:

$$(a - 2x)^7 = [a + (-2x)]^7$$

$$= a^7 + 7a^6(-2x) + \frac{7 \cdot 6}{2} a^5(-2x)^2 + \frac{7 \cdot 6 \cdot 5}{2 \cdot 3} a^4(-2x)^3$$

$$+ \frac{7 \cdot 6 \cdot 5 \cdot 4}{2 \cdot 3 \cdot 4} a^3(-2x)^4 + \frac{7 \cdot 6 \cdot 5 \cdot 4 \cdot 3}{2 \cdot 3 \cdot 4 \cdot 5} a^2(-2x)^5$$

$$+ \frac{7 \cdot 6 \cdot 5 \cdot 4 \cdot 3 \cdot 2}{2 \cdot 3 \cdot 4 \cdot 5 \quad 6} a^1(-2x)^6 + \frac{7 \cdot 6 \cdot 5 \cdot 4 \cdot 3 \cdot 2 \cdot 1}{2 \cdot 3 \cdot 4 \cdot 5 \cdot 6 \cdot 7} a^0(-2x)^7$$

$$(a - 2x)^7 = a^7 - 14a^6x + 84a^5x^2 - 280a^4x^3 + 560a^3x^4$$

$$- 672a^2x^5 + 448ax^6 - 128x^7.$$

Pascal's Triangle

The coefficients of $(a + b)^0$, $(a + b)^1$, $(a + b)^2$, ..., $(a + b)^n$ can be obtained from Pascal's Triangle:

Table 16.1

$(a + b)^0$						1					
$(a + b)^1$					1		1				
$(a + b)^2$				1		2		1			
$(a + b)^3$			1		3		3		1		
$(a + b)^4$		1		4		6		4		1	
$(a + b)^5$	1		5	10		10	5		1		
$(a + b)^6$	1	6	15	20	15	6	1				
$(a + b)^7$	1	7	21	35	35	21	7	1			
$(a + b)^8$	1	8	28	56	70	56	28	8	1		
$(a + b)^9$	1	9	36	84	126	126	84	36	9	1	

where each number in the triangle is the sum of the two numbers above it, or one if it is on the edge.

CHAPTER 17

PARTIAL FRACTIONS

If the degree of the numerator of a polynomial fraction is less than that of the denominator, then the fraction is said to be proper; otherwise it is said to be improper.

To decompose a given proper fraction into partial fractions is to resolve the fraction into a sum of simpler proper fractions.

Let $f(x)$ and $g(x)$ be polynomials and the degree of $f(x)$ be less than that of $g(x)$. To decompose $\dfrac{f(x)}{g(x)}$ into partial fractions is to find

$$\frac{f(x)}{g(x)} = p_1 + p_2 + \ldots + p_r$$

where each p_i has one of the forms

$$\frac{A}{(ux + v)^m} \quad \text{or} \quad \frac{Bx + C}{(ax^2 + bx + c)^n}$$

where $b^2 - 4ac < 0$, that is $ax^2 + bx + c$ is irreducible, and m, n are non-negative integers.

The method for decomposing a given rational fraction is given below.

Step 1: If the degree of the numerator is greater than that of denominator, rearrange the given fraction so that it is expressed as a sum of a polynomial and a proper rational fraction. (This can often be done through long division.)

Step 2: Express the denominator of the proper rational fraction as a product of different factors of the form $(ux + v)^m$ and/or $(ax^2 + bx + c)^n$ where $ax^2 + bx + c$ is irreducible and m, n are non-negative integers.

Step 3: For each factor of the form $(ux + v)^m$, $m \geq 1$, the decomposition of the proper rational fraction contains a sum of m partial fractions of the form

$$\frac{A_1}{ux + v} + \frac{A_2}{(ux + v)^2} + \cdots + \frac{A_m}{(ux + v)^m}$$

where each A_i is a real constant to be found later.

For each factor of the form $(ax^2 + bx + c)^n$, $n \geq 1$ and $b^2 - 4ac < 0$, the decomposition of the proper rational fraction contains a sum of n partial fractions of the form

$$\frac{A_1 x + B_1}{ax^2 + bx + c} + \frac{A_2 x + B_2}{(ax^2 + bx + c)^2} +$$

$$\cdots + \frac{A_n x + B_n}{(ax^2 + bx + c)^n}$$

where each A_i and B_i are real constants to be found in the next step.

Therefore, the proper rational fraction $\frac{f(x)}{g(x)}$ is now in the form

$$\frac{f(x)}{g(x)} = \underbrace{p_1 + p_2 + \cdots + p_r}_{\text{partial fractions}}$$

Step 4: Find the common denominator of $p_1, p_2, \ldots, p_r$.
Then express each of the partial fractions obtained in Step 3 as fractions $p_1^*, p_2^*, p_3^*, \ldots, p_r^*$ all having a common denominator $q(x)$. Thus we have,

$$\frac{f(x)}{g(x)} = p_1 + p_2 + \cdots + p_r = \frac{f^*(x)}{q(x)}$$

114

where $g(x) = q(x)$ and $f^*(x)$ is the sum of the numerators of $p_1^*, p_2^*, \ldots, p_r^*$. Using the identity $f(x) = f^*(x)$ we can obtain a system of equations which enables us to solve for $A_1, A_2, \ldots, A_m, A_n, B_1, B_2, \ldots B_m, \ldots$ etc. as illustrated by the following example.

Decompose

$$\frac{3x^2 + 2x - 2}{x^3 - 1}$$

into partial fractions.

<u>Solution</u>: The denominator can be factored into the product

$$x^3 - 1 = (x - 1)(x^2 + x + 1),$$

and we write:

$$\frac{3x^2 + 2x - 2}{x^3 - 1} = \frac{A}{x - 1} + \frac{Bx + C}{x^2 + x + 1}$$

$$= \frac{A(x^2 + x + 1) + (Bx + C)(x - 1)}{(x - 1)(x^2 + x + 1)}$$

Setting the numerators of the above fractions equal, we have:

$$3x^2 + 2x - 2 = A(x^2 + x + 1) + (Bx + C)(x - 1)$$

Now we multiply out and collect like powers of x. We obtain:

$$3x^2 + 2x - 2 = (A + B)x^2 + (A - B + C)x + (A - C)$$

Equating coefficients of like powers of x, we obtain:

$$A + B = 3$$

$$A - B + C = 2$$

$$A - C = -2.$$

Solving for A, B, and C, we find $A = 1$, $B = 2$, and $C = 3$. Therefore we have:

$$\frac{3x^2 + 2x - 2}{x^3 - 1} = \frac{1}{x - 1} + \frac{2x + 3}{x^2 + x + 1}$$

CHAPTER 18

COMPLEX NUMBERS

A complex number is an expression of the form a + bi, where a and b are real numbers and i = $\sqrt{-1}$. In the complex number, a + bi, a is called the real part and bi the imaginary part. The conjugate of a complex number a + bi is a-bi.

Algebraic operations with complex numbers:

A) To add/subtract complex numbers, add/subtract the real and imaginary parts separately.

e.g. (a + bi) + (a - bi) = 2a.

B) To multiply two complex numbers, treat the numbers as ordinary binomials and replace i² by -1.

e.g. (a + bi)(a - bi) = a² + b².

C) To divide two complex numbers, multiply the numerator and denominator of the fraction by the conjugate of the denominator, replacing i² by -1.

$$\frac{6 + 3i}{2 + 4i} = \frac{6 + 3i}{2 + 4i} \cdot \frac{2 - 4i}{2 - 4i}$$

$$= \frac{(6 + 3i)(2 - 4i)}{(2 + 4i)(2 - 4i)}$$

$$= \frac{12 + 6i - 24i - 12i^2}{4 + 8i - 8i - 16i^2}$$

$$= \frac{12 - 18i - 12(-1)}{4 - 16(-1)}$$

$$= \frac{12 - 18i + 12}{4 + 16}$$

$$= \frac{24 - 18i}{20}$$

$$= \frac{2(12 - 9i)}{2(10)}$$

$$= \frac{12 - 9i}{10}$$

$$= \frac{12}{10} - \frac{9}{10}i = \frac{6}{5} - \frac{9}{10}i$$

Complex numbers can be represented graphically. For example, the complex number x + yi can be represented graphically by the point P with the rectangular coordinates (x,y).

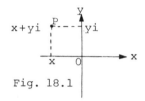

Fig. 18.1

Here, the x-axis is called the real axis since all points on the x-axis have coordinates of the form (x,0) and correspond to real numbers x + 0i = x. Similarly, the y-axis is called the imaginary axis since all points on the y-axis correspond to pure imaginary numbers 0 + yi = yi. The plane on which the complex numbers are represented is called the complex plane.

The complex numbers can also be represented by position vectors as illustrated in the following figure.

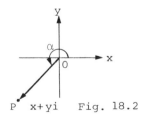

Fig. 18.2

The vector in the above figure has a magnitude A equal to $\sqrt{x^2 + y^2}$ and the direction of this vector is given by an angle $\alpha = \tan^{-1} \frac{y}{x}$. Hence, we may express this complex number $x + yi$ as the vector $A \leq \alpha$ where $A = \sqrt{x^2 + y^2}$, $\alpha = \tan^{-1} \frac{y}{x}$ and $A \leq \alpha = x + yi$.

The complex number $N = A(\cos\alpha + i\sin\alpha)$ is said to be in trigonometric (or polar) form, whereas the complex number $N = x + yi$ is said to be in rectangular form. Furthermore $A = \sqrt{x^2 + y^2}$ and $\alpha = \tan^{-1} \frac{y}{x}$.

DeMoivre's Theorem

The nth power of $A(\cos\theta + i\sin\theta)$ is given by:

$$[A(\cos\theta + i\sin\theta)]^n = A^n(\cos n\theta + i\sin n\theta).$$

CHAPTER 19

TRIGONOMETRY AND TRIGONOMETRIC FUNCTIONS

19.1 ANGLES AND TRIANGLES

An angle is the union of two rays having the same endpoint. The common endpoint is called the vertex of the angle.

An acute angle is an angle that is larger than 0° but smaller than 90°.

An angle whose measure is exactly 90° is called a right angle.

An obtuse angle is an angle that is larger than 90° but less than 180°.

An angle whose measure is exactly 180° is called a straight angle. Note: such an angle is in fact a straight line.

An angle that is greater than 180° but less than 360° is called a reflex angle.

Two angles whose sum is 180° are called supplementary angles.

Two angles whose sum is 90° are called complementary angles.

Two angles are called adjacent angles if and only if they have a common vertex and a common side lying between them.

A pair of nonadjacent angles with common vertex by two intersecting lines are called a pair of vertical angles. Vertical angles are equal.

Congruent angles are angles having equal measure.

A radian is defined as the measure of an angle, which when placed at the center of a circle, intercepts an arc of the circle equal to the radius of the circle.

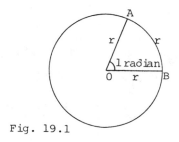

Fig. 19.1

In the figure above $\angle AOB = 1$ radian

$$1 \text{ radian} = \frac{180°}{\pi} \cong 57.3°.$$

$$1° = \frac{\pi}{180} \cong 0.0175 \text{ radian.}$$

A closed three-sided geometric figure is called a triangle.

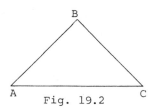

Fig. 19.2

A triangle with one right angle is called a right triangle. The side opposite the right angle in a right triangle is called the hypotenuse of the right triangle. The other two sides

120

are called the legs of the right triangle.

Fig. 19.3

A triangle that does not contain a right angle is called an oblique triangle.

The sum of the interior angles of a triangle is 180°.

A triangle can have at most one right or obtuse angle.

If a triangle has two equal angles, then the sides opposite those angles are equal.

If two sides of a triangle are equal, then the angles opposite those sides are equal.

The sum of the exterior angles of a triangle, taking one angle at each vertex is 360°.

A line that bisects one side of a triangle and is parallel to a second side, bisects the third side.

In a right triangle, the square of the hypotenuse is equal to the sum of the squares of the other two sides. This is commonly known as the theorem of Pythagoras or the Pythagorean theorm.

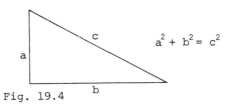

$$a^2 + b^2 = c^2$$

Fig. 19.4

A portion of a circle is called an arc of the circle.

An angle whose vertex is at the center of a circle and

whose sides are radii is called a central angle.

If α is the central angle in radians and r is the length of the radius of the circle, then the length of the intercepted arc 1 is given by 1 = αr.

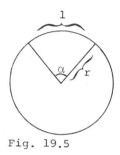

Fig. 19.5

19.2 TRIGONOMETRIC RATIOS

Given a right triangle △ABC as shown in Fig. 19.6

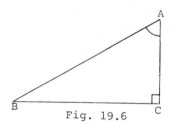

Fig. 19.6

Definition 1: $\sin A = \dfrac{BC}{AB}$

$$= \frac{\text{measure of side opposite } \angle A}{\text{measure of hypotenuse}}$$

Definition 2: $\cos A = \dfrac{AC}{AB}$

$$= \frac{\text{measure of side adjacent to } \angle A}{\text{measure of hypotenuse}}$$

Definition 3: $\tan A = \dfrac{BC}{AC}$

$$= \dfrac{\text{measure of side opposite } \angle A}{\text{measure of side adjacent to } \angle A}$$

Definition 4: $\cot A = \dfrac{AC}{BC}$

$$= \dfrac{\text{measure of side adjacent to } \angle A}{\text{measure of side opposite } \angle A}$$

$\sec A = \dfrac{AB}{AC}$

$$= \dfrac{\text{measure of hypotenuse}}{\text{measure of side adjacent to } \angle A}$$

$\csc A = \dfrac{AB}{BC}$

$$= \dfrac{\text{measure of hypotenuse}}{\text{measure of side opposite } \angle A}$$

The following table gives the values of sine, cosine, tangent and cotangent for some special angles.

Table 19.1

α	Sin α	Cos α	Tan α	Cot α
$0°$	0	1	0	∞
$\dfrac{\pi^R}{6} = 30°$	$\dfrac{1}{2}$	$\dfrac{\sqrt{3}}{2}$	$\dfrac{1}{\sqrt{3}}$	$\sqrt{3}$
$\dfrac{\pi^R}{4} = 45°$	$\dfrac{1}{\sqrt{2}}$	$\dfrac{1}{\sqrt{2}}$	1	1
$\dfrac{\pi^R}{3} = 60°$	$\dfrac{\sqrt{3}}{2}$	$\dfrac{1}{2}$	$\sqrt{3}$	$\dfrac{1}{\sqrt{3}}$
$\dfrac{\pi^R}{2} = 90°$	1	0	∞	0

19.3 TRIGONOMETRIC FUNCTIONS

A circle with center located at the origin of the

rectangular coordinate axes and radius equal to one unit length is called a unit circle.

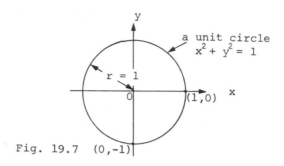

Fig. 19.7

An angle whose vertex is at the origin of a rectangular coordinate system and whose initial side coincides with the positive x-axis is said to be in standard position with respect to the coordinate system.

An angle in standard position with respect to a cartesian coordinate system whose terminal side lies in the first (or second or third or fourth) quadrant is called a first (or second or third or fourth) quadrant angle.

A quadrant angle is an angle in standard position whose terminal side lies on one of the axes of a cartesian coordinate system.

If θ is a non-quadrantal angle in standard position and $P(x,y)$ is any point, distinct from the origin, on the terminal side of θ, then the six trigonometric functions of θ are defined in terms of the abscissa (x-coordinate), ordinate (y-coordinate) and distance $\overline{OP}$ as follows:

$$\text{sine}\,\theta = \sin\theta = \frac{\text{ordinate}}{\text{distance}} = \frac{y}{r}$$

$$\text{cosine}\,\theta = \cos\theta = \frac{\text{abscissa}}{\text{distance}} = \frac{x}{r}$$

$$\text{tangent}\,\theta = \tan\theta = \frac{\text{ordinate}}{\text{abscissa}} = \frac{y}{x}$$

$$\text{cotangent}\,\theta = \cot\theta = \frac{\text{abscissa}}{\text{ordinate}} = \frac{x}{y}$$

$$\text{secant } \theta = \sec \theta = \frac{\text{distance}}{\text{abscissa}} = \frac{r}{x}$$

$$\text{cosecant} \theta = \csc \theta = \frac{\text{distance}}{\text{ordinate}} = \frac{r}{y}$$

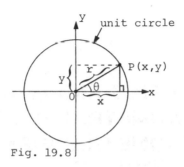

Fig. 19.8

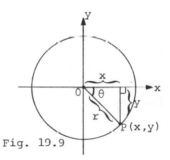

Fig. 19.9

The value of trigonometric functions of quadrantal angles are given in the table below.

Table 19.2

θ	$\sin\theta$	$\cos\theta$	$\tan\theta$	$\cot\theta$	$\sec\theta$	$\csc\theta$
0^0	0	1	0	$\pm\infty$	1	$\pm\infty$
90^0	1	0	$\pm\infty$	0	$\pm\infty$	1
180^0	0	-1	0	$\pm\infty$	-1	$\pm\infty$
270^0	-1	0	$\pm\infty$	0	$\pm\infty$	-1

The following table gives the signs of all the trigonometric functions for all four quadrants.

Table 19.3

Quadrant	sinα	cosα	tanα	cotα	secα	cscα
I	+	+	+	+	+	+
II	+	−	−	−	−	+
III	−	−	+	+	−	−
IV	−	+	−	−	+	−

19.4 PROPERTIES AND GRAPHS OF TRIGONOMETRIC FUNCTIONS; TRIGONOMETRIC IDENTITIES AND FORMULAS

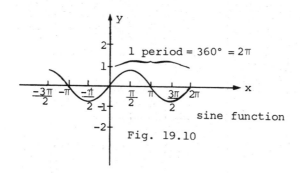

sine function

Fig. 19.10

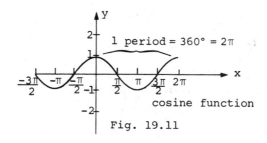

cosine function

Fig. 19.11

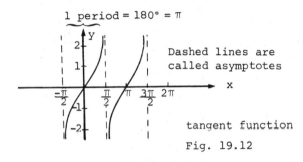

1 period = 180° = π

Dashed lines are
called asymptotes

tangent function

Fig. 19.12

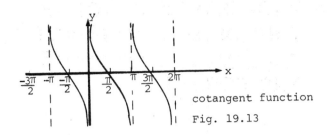

cotangent function

Fig. 19.13

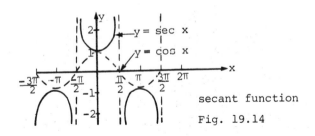

y = sec x

y = cos x

secant function

Fig. 19.14

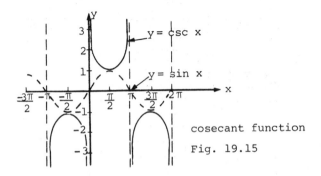

y = csc x

y = sin x

cosecant function

Fig. 19.15

127

A periodic function is defined as a function which repeats its values in definite cycles. Trigonometric functions are periodic functions.

For a function f(x), the smallest range of values of x which corresponds to a complete cycle of values of f(x) is called the period of the function, and it is denoted by T.

The frequency f of a periodic function f(x) with period T is frequency $f = \dfrac{1}{T}$.

The amplitude of a periodic function f(x) is the maximum value of its ordinate.

For a periodic function $f(x) = \sin(ax + \phi)$, ϕ is called the phase angle and $-\dfrac{\phi}{a}$ is called the phase shift.

Given $y = A \sin(Bx + C) + D$, the constant $|A|$ is the amplitude of the function, the constant B decides the period of the function, $T = \dfrac{2\pi}{|B|}$. C is the phase angle, and D will shift the graph of $y = A \sin(Bx + C)$ up (or down) along the y-axis by D units for positive (or negative) D.

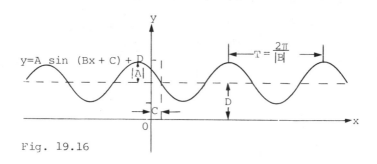

Fig. 19.16

One can combine two or more sine or cosine curves as illustrated by the following example to obtain more complicated forms of wave motions.

Graph (1) $y = 2 \sin x + 4 \sin x$

128

(2)   y = sin x + cos x

(1)   To obtain the graph of $y = 2 \sin x + 4 \sin x$ we first draw the graphs of $y_1 = 2 \sin x$ and $y_2 = 4 \sin x$ as shown in Fig. 19.17.   Next, draw the graph of $y = y_1 + y_2$ by adding corresponding ordinates as illustrated in Fig. 19.17. For instance, at $x = a_1$, the ordinate of y is $a_1 d_1$ which is the algebraic sum of the ordinates $a_1 b_1$ of y_1 and $a_1 c_1$ of y_2.

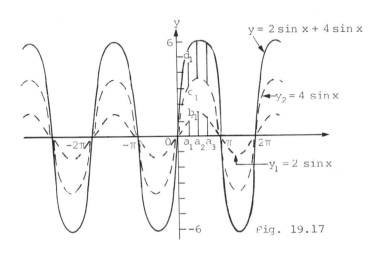

Fig. 19.17

Another way of obtaining the graph of the given function is to construct a table as the following.

Table 19.4

x	....	$-\frac{\pi}{2}$	$-\frac{\pi}{3}$	$-\frac{\pi}{4}$	$-\frac{\pi}{6}$	0	$\frac{\pi}{6}$	$\frac{\pi}{4}$	$\frac{\pi}{3}$	$\frac{\pi}{2}$	....
y=2sinx +4sinx	....	-6	$-3\sqrt{3}$	$-3\sqrt{2}$	-3	0	3	$3\sqrt{2}$	$3\sqrt{3}$	6	....

(2)   The graph of $y = \sin x + \cos x$ is obtained by using the methods described in part (1); it is shown in Fig. 19.18.

129

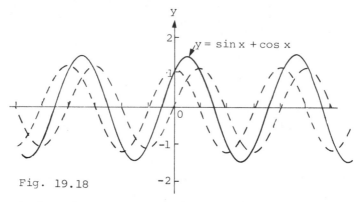

Fig. 19.18

A function f(x) is said to be an even function if f(-x) = f(x) for all independent variables x in the domain of f. f(x) is said to be an odd function if f(-x) = -f(x) for all x in the domain of f. Since for f(x) = sin x, f(-x) = sin(-x) = -sin x = -f(x), f(x) = sin x is an odd function.

On the other hand, f(x) = cos x is an even function since f(-x) = cos(-x) = cos x = f(x).

A curve C is said to be asymptotic to a straight line l if (a) the shortest distance d between a point on C and l is never zero, (b) d approaches zero as either x or y or both coordinates of P approaches ∞ or -∞. For example, one can clearly see that the curve of f(x) = tan x (Fig. 19.12) has the property of an asymptotic function.

We can represent the values of the trigonometric functions as line segments in a unit circle as follows.

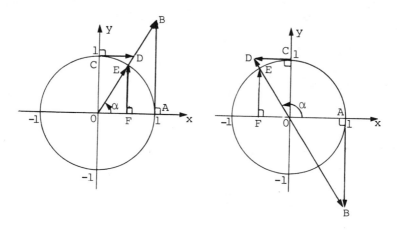

130

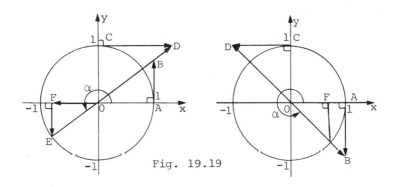

Fig. 19.19

$$\sin \alpha = \frac{FE}{OE} = FE$$

$$\cos \alpha = \frac{OF}{OE} = OF$$

$$\tan \alpha = \frac{FE}{OF} = \frac{AB}{OA} = AB$$

$$\cot \alpha = \frac{OF}{FE} = \frac{CD}{OC} = CD$$

$$\sec \alpha = \frac{OE}{OF} = \frac{OB}{OA} = OB$$

$$\csc \alpha = \frac{OE}{FE} = \frac{OD}{OC} = OD$$

Fundamental Relations and Identities

$$\sin^2\alpha + \cos^2 \alpha = 1$$

$$\tan \alpha = \frac{\sin \alpha}{\cos \alpha}$$

$$\cot\alpha = \frac{\cos \alpha}{\sin \alpha} = \frac{1}{\tan\alpha}$$

$$\csc \alpha = \frac{1}{\sin \alpha}$$

$$\sec \alpha = \frac{1}{\cos \alpha}$$

$$1 + \tan^2\alpha = \sec^2\alpha$$

$$1 + \cot^2\alpha = \csc^2\alpha$$

One can find all the trigonometric functions of an acute

131

angle when the value of any one of them is known.

For example, given α is an acute angle and $\csc\alpha = 2$. Then

$$\sin\alpha = \frac{1}{\csc\alpha} = \frac{1}{2}$$

$$\cos^2\alpha + \sin^2\alpha = 1, \quad \cos\alpha = \sqrt{1 - \sin^2\alpha}$$

$$= \sqrt{1 - (\tfrac{1}{2})^2}$$

$$= \sqrt{1 - \tfrac{1}{4}}$$

$$= \frac{\sqrt{3}}{2}$$

$$\tan\alpha = \frac{\sin\alpha}{\cos\alpha} = \frac{\tfrac{1}{2}}{\frac{\sqrt{3}}{2}} = \frac{1}{\sqrt{3}} = \frac{\sqrt{3}}{3}$$

$$\cot\alpha = \frac{1}{\tan\alpha} = \sqrt{3}$$

$$\sec\alpha = \frac{1}{\cos\alpha} = \frac{1}{\frac{\sqrt{3}}{2}} = \frac{2}{\sqrt{3}} = \frac{2\sqrt{3}}{3}$$

For a given angle θ in standard position, the related angle of θ is the unique acute angle which the terminal side of θ makes with the x-axis.

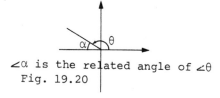

$\angle\alpha$ is the related angle of $\angle\theta$
Fig. 19.20

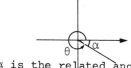

$\angle\alpha$ is the related angle of $\angle\theta$
Fig. 19.21

132

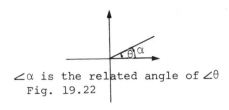

∠α is the related angle of ∠θ
Fig. 19.22

Let θ be an angle in standard position and φ be the related angle of θ.

A) If θ is a first quadrant angle, then

    a) $\sin\theta = \sin\phi$

    b) $\cos\theta = \cos\phi$

    c) $\tan\theta = \tan\phi$

    d) $\cot\theta = \cot\phi$

    e) $\sec\theta = \sec\phi$

    f) $\csc\theta = \csc\phi$

B) If θ is a second quadrant angle:

    a) $\sin\theta = \sin\phi$

    b) $\cos\theta = -\cos\phi$

    c) $\tan\theta = -\tan\phi$

    d) $\cot\theta = -\cot\phi$

    e) $\sec\theta = -\sec\phi$

    f) $\csc\theta = \csc\phi$

C) If θ is a third quadrant angle, then

    a) $\sin\theta = -\sin\phi$

    b) $\cos\theta = -\cos\phi$

    c) $\tan\theta = \tan\phi$

    d) $\cot\theta = \cot\phi$

    e) $\sec\theta = -\sec\phi$

    f) $\csc\theta = -\csc\phi$

D) If θ is a fourth quadrant angle, then

    a) $\sin \theta = -\sin \phi$

    b) $\cos \theta = \cos \phi$

    c) $\tan \theta = -\tan \phi$

    d) $\cot \theta = -\cot \phi$

    e) $\sec \theta = \sec \phi$

    f) $\csc \theta = -\csc \phi$

Table 19.5

	sin	cos	tan	cot	sec	csc
$-\alpha$	$-\sin \alpha$	$+\cos \alpha$	$-\tan \alpha$	$-\cot \alpha$	$+\sec \alpha$	$-\csc \alpha$
$90°+\alpha$	$+\cos \alpha$	$-\sin \alpha$	$-\cot \alpha$	$-\tan \alpha$	$-\csc \alpha$	$+\sec \alpha$
$90°-\alpha$	$+\cos \alpha$	$+\sin \alpha$	$+\cot \alpha$	$+\tan \alpha$	$+\csc \alpha$	$+\sec \alpha$
$180°+\alpha$	$-\sin \alpha$	$-\cos \alpha$	$+\tan \alpha$	$+\cot \alpha$	$-\sec \alpha$	$-\csc \alpha$
$180°-\alpha$	$+\sin \alpha$	$-\cos \alpha$	$-\tan \alpha$	$-\cot \alpha$	$-\sec \alpha$	$+\csc \alpha$
$270°+\alpha$	$-\cos \alpha$	$+\sin \alpha$	$-\cot \alpha$	$-\tan \alpha$	$+\csc \alpha$	$-\sec \alpha$
$270°-\alpha$	$-\cos \alpha$	$-\sin \alpha$	$+\cot \alpha$	$+\tan \alpha$	$-\csc \alpha$	$-\sec \alpha$
$360°+\alpha$	$+\sin \alpha$	$+\cos \alpha$	$+\tan \alpha$	$+\cot \alpha$	$+\sec \alpha$	$+\csc \alpha$
$360°-\alpha$	$-\sin \alpha$	$+\cos \alpha$	$-\tan \alpha$	$-\cot \alpha$	$+\sec \alpha$	$-\csc \alpha$

Addition and Subtraction Formulas

$$\sin(A \pm B) = \sin A \cos B \pm \cos A \sin B$$

$$\cos(A \pm B) = \cos A \cos B \mp \sin A \sin B$$

$$\tan(A \pm B) = \frac{\tan A \pm \tan B}{1 \mp \tan A \tan B}$$

$$\cot(A \pm B) = \frac{\cot A \cot B \mp 1}{\cot B \pm \cot A}$$

Double-angle Formulas

$$\sin 2A = 2\sin A \cos A$$

$$\cos 2A = 2\cos^2 A - 1$$

$$= 1 - 2\sin^2 A$$

$$= \cos^2 A - \sin^2 A$$

$$tan2A = \frac{2tanA}{1 - tan^2A}$$

Half-angle Formulas

$$\sin \frac{A}{2} = \pm \sqrt{\frac{1 - cosA}{2}}$$

$$\cos \frac{A}{2} = \pm \sqrt{\frac{1 + cosA}{2}}$$

$$\tan \frac{A}{2} = \pm \sqrt{\frac{1 - cosA}{1 + cosA}}$$

$$= \frac{1 - cosA}{sinA}$$

$$= \frac{sinA}{1 + cosA}$$

$$\cot \frac{A}{2} = \pm \sqrt{\frac{1 + cosA}{1 - cosA}} = \frac{1 + cosA}{sinA} = \frac{sinA}{1 - cosA}$$

Sum and Difference Formulas

$$\sin \alpha + \sin \beta = 2\sin\left(\frac{\alpha + \beta}{2}\right) \cos\left(\frac{\alpha - \beta}{2}\right)$$

$$\sin \alpha - \sin \beta = 2\cos\left(\frac{\alpha + \beta}{2}\right) \sin\left(\frac{\alpha - \beta}{2}\right)$$

$$\cos \alpha + \cos \beta = 2\cos\left(\frac{\alpha + \beta}{2}\right) \cos\left(\frac{\alpha - \beta}{2}\right)$$

$$\cos \alpha - \cos \beta = -2\sin\left(\frac{\alpha + \beta}{2}\right) \sin\left(\frac{\alpha - \beta}{2}\right)$$

$$\tan \alpha + \tan \beta = \frac{\sin(\alpha + \beta)}{\cos \alpha \cos\beta}$$

$$\tan \alpha \cdot \tan \beta = \frac{\sin(\alpha - \beta)}{\cos \alpha \cos\beta}$$

Product Formulas of Sines and Cosines

$$sinAsinB = \tfrac{1}{2}[\cos(A - B) - \cos(A + B)]$$

$$cosAcosB = \tfrac{1}{2}[\cos(A + B) + \cos(A - B)]$$

$$\sin A \cos B = \tfrac{1}{2}[\sin(A + B) + \sin(A - B)]$$

$$\cos A \sin B = \tfrac{1}{2}[\sin(A + B) - \sin(A - B)]$$

CHAPTER 20

SOLVING TRIANGLES

20.1 LAWS AND FORMULAS

The following results hold for any oblique triangle ABC with sides of length a, b, c opposite to vertices A, B, C respectively.

Law of Sines

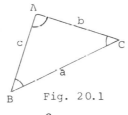

Fig. 20.1

$$\frac{a}{\sin A} = \frac{b}{\sin B} = \frac{c}{\sin C}$$

Law of Cosines

$$a^2 = b^2 + c^2 - 2bc \cos A.$$

$$b^2 = a^2 + c^2 - 2ac \cos B.$$

$$c^2 = a^2 + b^2 - 2ab \cos C.$$

Law of Tangents

$$\frac{a - b}{a + b} = \frac{\tan\left(\dfrac{A - B}{2}\right)}{\tan\left(\dfrac{A + B}{2}\right)}$$

$$\frac{b - c}{b + c} = \frac{\tan\left(\dfrac{B - C}{2}\right)}{\tan\left(\dfrac{B + C}{2}\right)}$$

$$\frac{a - c}{a + c} = \frac{\tan\left(\dfrac{A - C}{2}\right)}{\tan\left(\dfrac{A + C}{2}\right)}$$

Mollweide's Formulas

$$\frac{a + b}{c} = \frac{\cos \tfrac{1}{2}(A - B)}{\sin \dfrac{C}{2}}, \qquad \frac{a - b}{c} = \frac{\sin \tfrac{1}{2}(A - B)}{\cos \dfrac{C}{2}}$$

$$\frac{b + c}{a} = \frac{\cos \tfrac{1}{2}(B - C)}{\sin \dfrac{A}{2}}, \qquad \frac{b - c}{a} = \frac{\sin \tfrac{1}{2}(B - C)}{\cos \dfrac{A}{2}}$$

$$\frac{c + a}{b} = \frac{\cos \tfrac{1}{2}(C - A)}{\sin \dfrac{A}{2}}, \qquad \frac{c - a}{b} = \frac{\sin \tfrac{1}{2}(C - A)}{\cos \dfrac{B}{2}}$$

Projection Formulas

$$BC = b \cos C + c \cos B$$

$$AC = a \cos C + c \cos A$$

$$AB = a \cos B + b \cos A$$

In a 30°-60° right triangle, (Fig. 20.2), the hypotenuse is twice the length of the side opposite the 30° angle. The side opposite the 60° angle is equal to the length of the side opposite the 30° angle multiplied by $\sqrt{3}$.

In an isosceles 45° right triangle, (Fig. 20.3), the hypotenuse is equal to the length of one of its arms multiplied by $\sqrt{2}$.

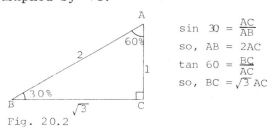

$$\sin 30 = \frac{AC}{AB}$$
so, $AB = 2AC$
$$\tan 60 = \frac{BC}{AC}$$
so, $BC = \sqrt{3}\, AC$

$$\sin 30° = \frac{AC}{AB}$$

so, $AB = 2AC$

$$\tan 60° = \frac{BC}{AC}$$

Fig. 20.2

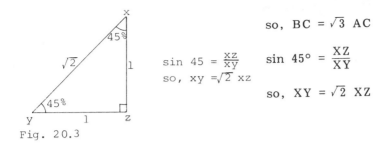

so, $BC = \sqrt{3}\ AC$

$\sin 45 = \frac{XZ}{XY}$ $\sin 45° = \frac{XZ}{XY}$

so, $xy = \sqrt{2}\ xz$

so, $XY = \sqrt{2}\ XZ$

Fig. 20.3

20.2 IMPORTANT CONCEPTS
AND THEOREMS

The altitude h on the hypotenuse of a right triangle is the mean proportional between the segments of the hypotenuse, also called the geometric mean.

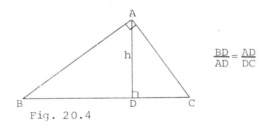

$\frac{BD}{AD} = \frac{AD}{DC}$

Fig. 20.4

In a right triangle $\triangle$ ABC, the altitude to the hypotenuse, $\overline{AD}$, separates the triangle into two triangles that are similar to each other and to the original triangle.

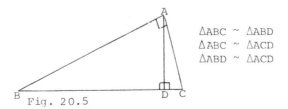

$\triangle ABC \sim \triangle ABD$
$\triangle ABC \sim \triangle ACD$
$\triangle ABD \sim \triangle ACD$

Fig. 20.5

The length of the median to the hypotenuse of a right

139

triangle is equal to one-half the length of the hypotenuse.

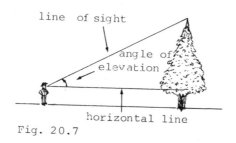

CD = 1/2 AB

Fig. 20.6

In solving triangles, terms such as line of sight, angle of elevation and angle of depression are often used. These terms are illustrated below.

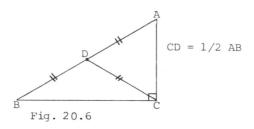

line of sight

angle of elevation

horizontal line

Fig. 20.7

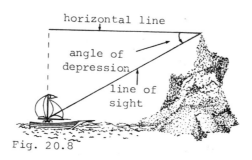

horizontal line

angle of depression

line of sight

Fig. 20.8

Ex. At a point on the ground 40 feet from the foot of a tree, the angle of elevation to the top of the tree is 42°. Find the height of the tree to the nearest foot.

140

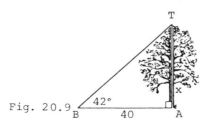

Fig. 20.9

T

42°

B   40   A

x

Solution: The geometric figure formed is a right triangle (see figure). Since the unknown height of the tree is opposite the given angle of elevation, and we are given the side adjacent to this angle, we can solve the problem using the tangent ratio. The tangent is the ratio of the length of the leg opposite the acute angle to the length of the leg adjacent to the acute angle in any right triangle. In this example,

$$\tan B = \frac{\text{length of leg opposite} \quad \angle B}{\text{lenght of leg adjacent} \quad \angle B},$$

$$\tan B = \frac{AT}{BA}.$$

Let x = AT, and consult a standard table of tangents to find that tan 42° = 0.9004. Since BA = 40, we obtain

$$0.9004 = \frac{x}{40}.$$

Therefore,   x = 40(0.9004) = 36.016.

Therefore, the height of the tree, to the nearest foot, is 36 feet.

The following theorems are often used for finding the area of a triangle.

Theorem: The area of a triangle is given by A = ½bh, where b is the length of the base and h is the perpendicular height of the triangle.

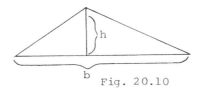

h

b   Fig. 20.10

Theorem: The area of a triangle equals one-half the product of any two adjacent sides and the sine of the included angle.

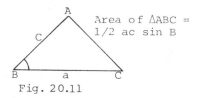

Area of $\triangle ABC =$
$1/2$ ac sin B

Fig. 20.11

Theorem: Triangles that share the same base and have their third vertex on a line parallel to the base, have equal areas.

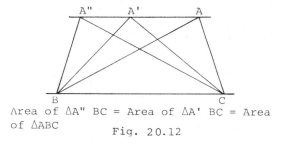

Area of $\triangle A''$ BC = Area of $\triangle A'$ BC = Area
of $\triangle ABC$

Fig. 20.12

The areas of two triangles having equal bases, have the same ratio as that of their altitudes and vice versa.

The area of a triangle, the length of whose three sides are a, b, and c, is given by the formula

$$A = \sqrt{s(s - a)(s - b)(s - c)}$$

where $s = \frac{1}{2}(a + b + c)$; the semiperimeter of the triangle. The above formula is commonly referred to as Heron's formula.

The area of an equilateral triangle is given by the formula,

$$A = \frac{x^2 \sqrt{3}}{4},$$

where x is the length of a side of the triangle.

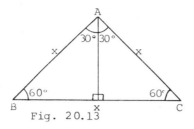

Fig. 20.13

Theorem: The area of an equilateral triangle equals $\frac{\sqrt{3}}{3}$ times the square of the altitude of the triangle.

$AP = h$

Area of $\triangle ABC = \frac{\sqrt{3}}{3} h^2$

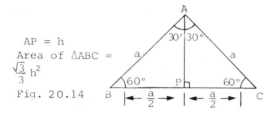

Fig. 20.14

Theorem: A median drawn to a side of a triangle divides the triangle into two triangles of equal area.

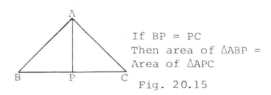

If $BP = PC$
Then area of $\triangle ABP =$
Area of $\triangle APC$

Fig. 20.15

Theorem: The area of an isosceles triangle whose congruent sides have length l, with included angle α is:.

$$A = \tfrac{1}{2}l^2 \sin\alpha.$$

Area is also given by the formula

$$A = h^2 \tan \frac{\alpha}{2} ;$$

where h is the length of the altitude to the side opposite to the angle α.

143

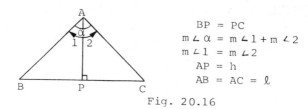

$$BP = PC$$
$$m \angle \alpha = m \angle 1 + m \angle 2$$
$$m \angle 1 = m \angle 2$$
$$AP = h$$
$$AB = AC = \ell$$

Fig. 20.16

CHAPTER 21

INVERSE TRIGONOMETRIC FUNCTIONS AND TRIGONOMETRIC EQUATIONS

21.1 INVERSE TRIGONOMETRIC FUNCTIONS

By definition, if for every number y in the range of a function f there is exactly one number x in the domain of f such that y = f(x), then f is one-to-one. Then f has an inverse function f^{-1} whose domain is the range of f and whose range is f's domain.

If the domain of the six trigonometric functions is not restricted, then they have no inverses, because they repeat and thus are not one-to-one. If, however, the domains are restricted to suitable intervals, we grant ourselves one-to-one correspondence, and we can define an inverse for the intervals.

Here are the trigonometric functions and their inverses:

If   y = sin x,   then $x = \sin^{-1} y = $ Arcsin y

     y = cos x, $x = \cos^{-1} y = $ Arccos y

     y = tan x, $x = \tan^{-1} y = $ Arctan y.

The others are written the same way.

Definition of Inverse Trigonometric Functions

1. Sine^{-1}

$T_1 = \{(y,x) \mid (x,y) \text{ satisfies } y = \sin x, -\frac{\pi}{2} \le x \le \frac{\pi}{2}\}$

2. Cosine^{-1}

$T_2 = \{(y,x) \mid (x,y) \text{ satisfies } y = \cos x,\ 0 \le x \le \pi\}$

3. Tangent^{-1}

$T_3 = \{(y,x) \mid (x,y) \text{ satisfies } y = \tan x,\ \frac{-\pi}{2} < x < \frac{\pi}{2}\}$

4. Cotangent^{-1}

$T_4 = \{(y,x) \mid (x,y) \text{ satisfies } y = \cot x,\ 0 < x < \pi\}$

5. Secant^{-1}

$T_5 = \{(y,x) \mid (x,y) \text{ satisfies } y = \sec x,\ 0 \le x < \frac{\pi}{2},\ \frac{\pi}{2} < x \le \pi\}$

6. Cosecant^{-1}

$T_6 = \{(y,x) \mid (x,y) \text{ satisfies } y = \csc x,\ \frac{-\pi}{2} \le x < 0,\ 0 < x \le \frac{\pi}{2}\}$

Capital letters at the beginning indicate the functions defined above.

The graph of the standard inverse trigonometric functions are shown by solid portions of the curves given below.

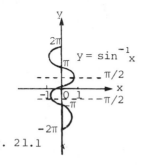

Fig. 21.1

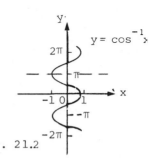

Fig. 21.2

146

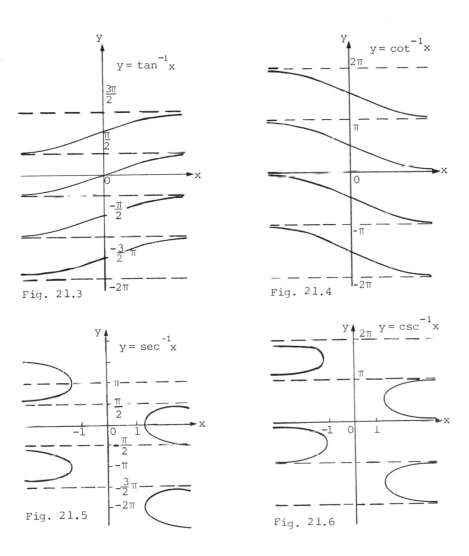

Fig. 21.3 $y = \tan^{-1}x$

Fig. 21.4 $y = \cot^{-1}x$

Fig. 21.5 $y = \sec^{-1}x$

Fig. 21.6 $y = \csc^{-1}x$

21.2 TRIGONOMETRIC EQUATIONS

A trigonometric equation is an equation which involves one or more trigonometric functions of an unknown angle.

Trigonometric equations are of two types – the identity

and the conditional equation.

An identity is a true statement for all values in the replacement set for which the equation is defined.

ex. $\sin^2\theta + \cos^2\theta = 1$

A conditional equation is a true statement only for particular values or sets of the variable quantities involved.

ex. $\sin\theta + \cos\theta = 1$

Ex. Solve the equation

$$\sin^2\theta + 2\cos\theta - 1 = 0$$

for non-negative values of θ less than 2π.

Solution: Two trigonometric functions of the unknown θ itself appear in this equation. Accordingly, we make use of the identity connecting these functions, namely,

$$\sin^2\theta + \cos^2\theta = 1,$$

to transform it into an equation involving only one function of θ. Replace $\sin^2$ by $1 - \cos^2\theta$.

$$\sin^2\theta + 2\cos\theta - 1 = 0$$

$$1 - \cos^2\theta + 2\cos\theta - 1 = 0.$$

Factor out $\cos\theta$

$$\cos\theta(2 - \cos\theta) = 0.$$

Whenever a product of two numbers $ab = 0$, either $a = 0$ or $b = 0$, hence $\cos\theta = 0$ or $2 - \cos\theta = 0$. Thus $\cos\theta = 0$ or $\cos\theta = 2$.

Now there are two angles in the range $0 \leqq \theta < 2\pi$ for which $\cos\theta = 0$ namely

$$\theta = \frac{\pi}{2}, \qquad \theta = \frac{3\pi}{2}.$$

But, since a cosine of an angle can never exceed untiy, the relation $\cos\theta = 2$ does not yield a value of θ. Hence we have just two solutions, as given above. It is easy to

148

check these solutions.

Check: $\sin^2 \theta + 2 \cos \theta - 1 = 0$

$(\sin \theta)^2 + 2 \cos \theta - 1 = 0.$

For $\theta = \dfrac{\pi}{2}$

$(\sin \dfrac{\pi}{2})^2 + 2 \cos \dfrac{\pi}{2} - 1 = 0$

$1 + 2 \cdot 0 - 1 = 0$
$0 = 0 \checkmark.$

For $\theta = \dfrac{3}{2} \pi$

$(\sin \dfrac{3}{2} \pi)^2 + 2 \cos \dfrac{3}{2} \pi - 1 = 0$

$(-1)^2 + 2 \cdot 0 - 1 = 0$
$0 = 0 \checkmark.$

CHAPTER 22

INTRODUCTION TO ANALYTIC GEOMETRY

Analytic geometry refers to the study of geometric figures using algebraic principles. This approach, commonly referred to as coordinate geometry has been accredited to a seventeenth century French mathematician, René Descartes.

Postulate: The points on a straight line can be placed in a one-to-one correspondence with real numbers such that for every point of the line there corresponds a unique real number, and for every real number there corresponds a unique point of the line.

A number scale is a straight line on which distances from a point are numbered in equal units; positively in one direction and negatively in the opposite direction. The origin is the zero point from which distances are measured.

The Cartesian Product of a set X and a set Y is the set of all ordered pairs (x,y) where x belongs to X and y belongs to Y.

The graph of $\mathbb{R} \times \mathbb{R}$ is called the Cartesian coordinate plane. Graphically, it consists of a pair of perpendicular lines called coordinate axes, and the plane they lie in. The vertical axis is the y-axis and the horizontal axis is the x-axis. The point of intersection of these two axes is called the origin. It is the zero point of both axes. Furthermore, points to the right of the origin on the x-axis and above the origin on the y-axis represents positive real numbers. The negative numbers are represented similarly. Each element of the set $\mathbb{R} \times \mathbb{R}$ is represented by a point on the Cartesian coordinate plane.

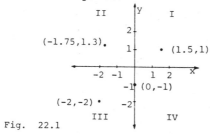

Fig. 22.1

150

The abscissa is the x-coordinate of a given point, and the ordinate is the y-coordinate of the given point.

The plane is divided into four regions by the coordinate axes. These regions are numbered from one to four starting from the region where both x and y values are positive and proceeding counterclockwise.

The distance between any two points on a number scale is the absolute value of the difference between the corresponding numbers.

Theorem 1: For any two points A and B with coordinates (x_A, y_A) and (x_B, y_B) respectively, the distance between A and B is

$$d(A,B) = \sqrt{(x_A - x_B)^2 + (y_A - y_B)^2},$$

this is commonly known as the distance formula.

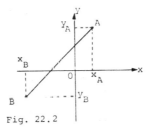

Fig. 22.2

Theorem 2: Given a line segment with endpoints (x_A, y_A) and (x_B, y_B), the coordinates of the midpoint of the line segment are (x_m, y_m) where

$$x_m = \frac{x_A + x_B}{2}, \qquad y_m = \frac{y_A + y_B}{2}$$

this is commonly known as the midpoint formula.

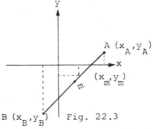

Fig. 22.3

The distance d between a given point A with coordinates (x_A, y_A) and a line 1 defined as $ax + by + c = 0$ is given by the following formula:

$$d = \left| \frac{ax_A + by_A + c}{\sqrt{a^2 + b^2}} \right|$$

A parabola is the locus of points whose distance from a fixed line, called the directrix and a fixed point, called the focus is equal.

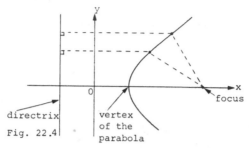

directrix

Fig. 22.4

vertex
of the
parabola

focus

An ellipse is the locus of points, the sum of whose distances from two fixed points (foci) is a constant.

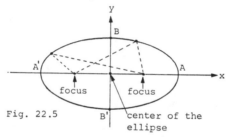

Fig. 22.5

focus   focus

center of the
ellipse

The line segment BB' is called the minor axis and the line segment AA' is called the major axis.

A hyperbola is the locus of all points the difference of whose distance from two fixed points is a constant.

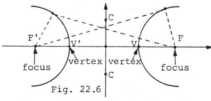

focus   vertex vertex   focus

Fig. 22.6

The line segment VV' joining the two vertices is the hyperbola's transverse axis and the segment CC' is the conjugate axis.

The equation of a circle centered at (x_0, y_0) with radius r is given by: $(x - x_0)^2 + (y - y_0)^2 = r^2$, where x_0 and y_0 are the coordinates of the center of the circle and r is the length of the radius.

152

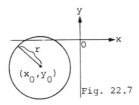

Fig. 22.7

The equation of a parabola with vertex V at (x_0, y_0) and directrix d at $x = x_0 - p$ is

$$(y - y_0)^2 = 4P(x - x_0)$$

Note that the focus is then at $(x_0 + p, y_0)$.

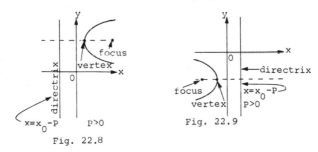

Fig. 22.8

Fig. 22.9

The equation of a parabola with vertex, $V(x_0, y_0)$ and directrix d, $y = y_0 - p$ is

$$(x - x_0)^2 = 4P(y - y_0)$$

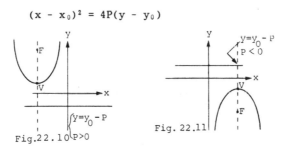

Fig. 22.10

Fig. 22.11

Theorem: The equation of an ellipse with center $C(x_0, y_0)$ and the major axes parallel respectively to the x-axis and y-axis are given below:

A) $\dfrac{(x - x_0)^2}{a^2} + \dfrac{(y - y_0)^2}{b^2} = 1, \quad a \geq b > 0$

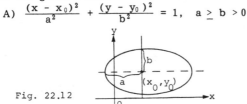

Fig. 22.12

153

B) $\dfrac{(y - y_0)^2}{a^2} + \dfrac{(x - x_0)^2}{b^2} = 1, \quad a \geq b > 0$

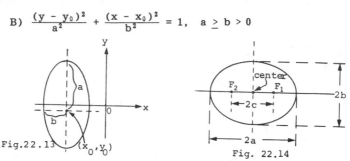

Fig.22.13

Fig. 22.14

where the length of the major axis is 2a and the length of the minor axis is 2b. $c = \sqrt{a^2 - b^2}$ is half the distance between the two foci.

Theorem: The equation of a hyperbola with center $C(x_0, y_0)$, with transverse axis parallel to the x-axis is:

$$\dfrac{(x - x_0)^2}{a^2} - \dfrac{(y - y_0)^2}{b^2} = 1$$

where the length of the transverse axis is 2a and the length of the conjugate axis is 2b.

$c = \sqrt{a^2 + b^2}$ is half the distance between the two foci F_1 and F_2.

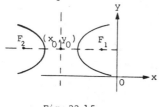

Fig. 22.15

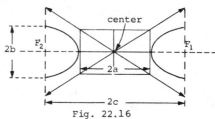

Fig. 22.16

A point is expressed in polar form, (r, θ), as a distance and an angle instead of an x-coordinate and a y-coordinate. Draw a line from the point to the origin. r is the length of that line, θ is the angle the line makes positively (counterclockwise) with the positive x-axis.

For example $(5, 60°)$ represents the point 5 units from the origin and 60° around from the positive x-axis.

To change rectangular to polar coordinates:

$r = \pm\sqrt{x^2 + y^2}$

$\theta = \tan^{-1} \dfrac{y}{x}$

To change polar to rectangular coordintes:

$x = r \cos \theta$

$y = r \sin \theta$

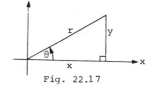

Fig. 22.17